YOUR KNOWLEDGE HAS

Bibliographic information published by the German National Library:

The German National Library lists this publication in the National Bibliography; detailed bibliographic data are available on the Internet at http://dnb.dnb.de .

Imprint:

Copyright © 2015 GRIN Verlag, Open Publishing GmbH
Print and binding: Books on Demand GmbH, Norderstedt Germany
ISBN: 9783668555457

This book at GRIN:

http://www.grin.com/en/e-book/378120/understanding-the-acceptance-of-3d-printing-toolkits-an-extension-of-the

Johannes Köck

Understanding the acceptance of 3D printing toolkits.
An extension of the technology acceptance model

GRIN Publishing

GRIN - Your knowledge has value

Since its foundation in 1998, GRIN has specialized in publishing academic texts by students, college teachers and other academics as e-book and printed book. The website www.grin.com is an ideal platform for presenting term papers, final papers, scientific essays, dissertations and specialist books.

Visit us on the internet:

http://www.grin.com/

http://www.facebook.com/grincom

http://www.twitter.com/grin_com

Innovation Technology

"Understanding the acceptance of 3D printing toolkits: An extension of the technology acceptance model"

Submitted by:

Name: Johannes Köck

1 Introduction

1.1 Problem statement

Additive manufacturing, also known as 3D printing (3DP), is a technology which gained a lot of interest in recent years. The market is supposed to grow further with a new annual growth record of 35 % in 2013. However, the world leading market report for additive manufacturing, the Wohlers report, states that growth in the upcoming years is especially going to be driven by "3-D printers that cost less than $5,000, as well as the expanded use of the technology for the production of parts, especially metal, that go into final products." [1] Consequently, the following paper focuses on 3D printing for non-experts as more and more citizens can afford this technology and as there is not a lot of research in the field of information systems about 3D printing on the consumer level. Questions such as "What are the needs of consumers regarding 3D printing? Which are the top products the consumers want to produce? How do these non-experts deal with 3D printing design software?" have not been answered satisfactory yet.

1.2 Research objective

The goal of this paper to find out factors that affect the acceptance of 3D printing toolkits. A hypothesized research model for 3D printing toolkits is proposed. Based on a survey of 30 participants this research model is analyzed and evaluated. The result is, that five of these seven proposed determinants have a strong influence on the Behavioral Intention to Use such a toolkit.

1.3 Research paradigm in Information Systems

There are two different fields of IS research: Behavorial science (BS) and Design Science (DS) [2] [3]. For both fields an effective literature review is the basis. However, while the design-oriented research creates IT artifacts the behavioral research creates theories about these artifacts and tries to check the truth of these theories. Theories that have been found to be empirically adequate, in turn, serve design-oriented researchers for new IT artifacts [2].

Consequently, as this research is based on TAM and as it tries to validate the truth of TAM related to 3DP, this research paradigm is in the field of BS. The validation of this research is going to take place with a study/survey and a subsequent evaluation of the gained data.

2   Research model

The Technology Acceptance Model (TAM) is an information system theory. This model was developed by Fred Davis in his dissertation of 1986 [4]. Since, this model has evolved to one of the most cited models in the area of technology diffusion [5]. Although there are many models which predict the use of a system TAM has gained most attention in information systems [6].

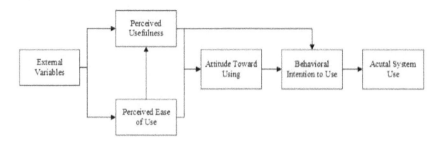

Figure 1: Technology Acceptance Model according to [7]

Figure 1 shows the original TAM proposed in 1989 by Davis, Bagozzi and Warshaw. In 2000, Davis and Venkatesh introduced TAM2, which is an extension of TAM. This model explains usefulness and intentions to use a system in terms of "social influence and cognitive instrumental processes" [8]. However, this theory was further adapted in 2003 by Venkatesh, Davis and Morris as a Unified Theory of Acceptance and Use of Technology (UTAUT) and in 2008 by Venkatesh and Bala as Technology Acceptance Model 3 (TAM3) [9, 10] .

However, integrating literature and hypotheses mentioned in the next section the proposed research model for this current study is shown in Figure 2.

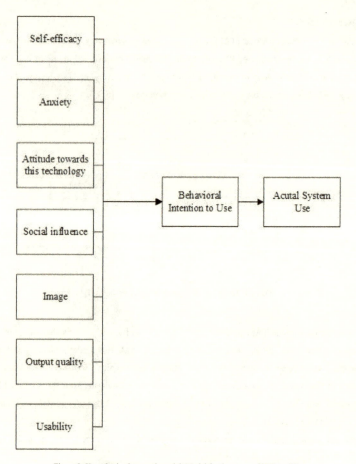

Figure 2: Hypothesized research model: Model for the usage of 3DP toolkits

3 Method

3.1 Measurement scales and items

In order to formulate respective questions for the intended survey existing questions from previous studies function as a basis. This section presents potential questions of previous research which might be suitable for this research. However, as no research paper related to TAM and 3DP has been published so far the potential questions for this study have to be adapted. These questions are based on a couple of determinants of TAM, UTAUT and TAM2.

Ease of Use

Study	Question
Davis, Fred D. (1989): Perceived Usefulness, Perceived Ease of Use, and User Acceptance of Information Technology. In MIS Quarterly 13 (3), pp. 319–340. DOI: 10.2307/249008.	Learning to operate X would be easy for me. I would find it easy to get X to do what I want to do. My interaction with X would be clear and understandable. I would find X to be flexible to interact with. It would be easy for me to become skillful at using X. I would find X easy to use.
Venkatesh, Viswanath; Davis, Fred D.; Davis, Gordon B.; Morris, Michael G. (2003): User Acceptance of Information Technology: Toward a Unified View. In MIS Quarterly 27 (3), pp. 425–478.	My interaction with the system would be clear and understandable. It would be easy for me to become skillful at using the system. I would find the system easy to use. Learning to operate the system is easy for me.
Venkatesh, Viswanath; Davis, Fred D. (2000): A Theoretical Extension of the Technology Acceptance Model: Four Longitudinal Field Studies. In Management Science 46 (2), pp. 186–204. DOI: 10.1287/mnsc.46.2.186.11926.	I find it easy to get the system to do what I want it to do. Interaction with the system is clear and understandable.

Usefulness

Davis, Fred D. (1989): Perceived Usefulness, Perceived Ease of Use, and User Acceptance of	I would find X useful.

Information Technology. In MIS Quarterly 13 (3), pp. 319–340. DOI: 10.2307/249008.	
Venkatesh, Viswanath; Davis, Fred D. (2000): A Theoretical Extension of the Technology Acceptance Model: Four Longitudinal Field Studies. In *Management Science* 46 (2), pp. 186–204. DOI: 10.1287/mnsc.46.2.186.11926.	Using the system enhances my effectiveness.

Anxiety

Venkatesh, Viswanath; Davis, Fred D.; Davis, Gordon B.; Morris, Michael G. (2003): User Acceptance of Information Technology: Toward a Unified View. In MIS Quarterly 27 (3), pp. 425–478.	I feel apprehensive about using the system. It scares me to think that I could lose a lot of information using the system by hitting the wrong key. I hesitate to use the system for fear of making mistakes I cannot correct. The system is somewhat intimidating to me.

Self-efficacy

Venkatesh, Viswanath; Davis, Fred D.; Davis, Gordon B.; Morris, Michael G. (2003): User Acceptance of Information Technology: Toward a Unified View. In MIS Quarterly 27 (3), pp. 425–478.	I could complete a job or task using the system… If there was no one around to tell me what to do as I go. If I could call someone for help if I got stuck. If I had a lot of time to complete the job for which the software was provided. If I had just the built-in help facility for assistance.

Social influence

Venkatesh, Viswanath; Davis, Fred D.; Davis, Gordon B.; Morris, Michael G. (2003): User Acceptance of Information Technology: Toward a Unified View. In *MIS Quarterly* 27 (3), pp. 425–478.	People who influence my behavior think that I should use the system. People who are important to me think that I should use the system.

Image

Venkatesh, Viswanath; Davis, Fred D. (2000): A Theoretical Extension of the Technology Acceptance Model: Four Longitudinal Field Studies. In *Management Science* 46 (2), pp. 186–204. DOI: 10.1287/mnsc.46.2.186.11926.	Having the system is a status symbol. People who use the system have more prestige than those who do not.

Voluntariness

Venkatesh, Viswanath; Davis, Fred D. (2000): A Theoretical Extension of the Technology Acceptance Model: Four Longitudinal Field Studies. In Management Science 46 (2), pp. 186–204. DOI: 10.1287/mnsc.46.2.186.11926.	My use of the system is voluntary.

Quality Output

Venkatesh, Viswanath; Davis, Fred D. (2000): A Theoretical Extension of the Technology Acceptance Model: Four Longitudinal Field Studies. In *Management Science* 46 (2), pp. 186–204. DOI: 10.1287/mnsc.46.2.186.11926.	The quality output I get from the system is high. I have no problem with the quality of the system's output.

Behavioral Intention to Use (Usage)

Venkatesh, Viswanath; Davis, Fred D.; Davis, Gordon B.; Morris, Michael G. (2003): User Acceptance of Information Technology: Toward a Unified View. In *MIS Quarterly* 27 (3), pp. 425–478.	I intend to use the system in the next X months. I predict I would use the system in the next X months. I plan to use the system in the next X months.

In the following respective measurement scales for this study are selected, adapted and explained. These are determinants which affect the *Behavioral Intention to Use*.

Self-efficacy:

"The degree to which an individual believes that he/she has the ability to perform a respective job by using a [toolkit]" [10]. This measurement scale has two items.

I could complete a job or task using the toolkit...
- If there was no one around to tell me what to do as I go.
- If I had a lot of time to complete the job for which the toolkit was provided.

Anxiety

"The degree of an individual's apprehension, or even fear, when he/she is faced with the possibility of using a [toolkit]" [10]. This scale consists of three items.
- I feel apprehensive about using the toolkit.
- It scares me to think that I could lose a lot of information using the toolkit by hitting the wrong key.
- I hesitate to use the toolkit for fear of making mistakes I cannot correct.

Attitude towards this technology (Usefulness) [11]

"The degree to which a person believes that using a particular system would enhance his or her job performance" [12]. This measurement scale has three items.
- I find the toolkit useful.
- Using the toolkit enhances my effectiveness.
- The toolkit supports to get my ideas into physical objects.

Quality Output

"The degree to which an individual believes that the system performs his or her job tasks well" [8]. This scale consists of two items.
- The quality output I get from the toolkit is high.
- I have no problem with the quality of the toolkit's output.

Image

"The degree to which an individual perceives that use of an innovation will enhance his or her status in his or her social system" [10]. This measurement scale has two items.
- Having the toolkit/3D printer is a status symbol.
- People who use the toolkit/3D printer have more prestige than those who do not.

Social influence

"The degree to which an individual perceives that most people who are important to him think he should or should not use the system" [8]. This measurement scale has two items.
- People who influence my behavior think that I should use the toolkit/3DP.

- People who are important to me think that I should use the toolkit/3DP.

Usability (Ease of Use) [11]

"The degree to which a person believes that using a [toolkit] will be free of effort" [7]. This scale consists of four items.

- Learning to deal with the toolkit was easy.
- I find it easy to achieve with the system exactly what I want.
- My interaction with the toolkit was clear and understandable.
- It would be easy for me to become skillful at using the toolkit.

Behavioral Intention to Use (Usage)

This scale consists of two items.

- I intend to use such a toolkit in the next 12 months.
- I am sure I am going to use a 3D printing toolkit in the upcoming years.

3.2   Study and survey

According to the presented scale measurements a questionnaire is designed which has to be filled out by participants subsequent to the study. The study itself consists of providing a free toolkit to non-experts. These non-experts have to create the same object. Previous studies found out top household objects which people want to produce in their homes by 3D printing [13]. Referring to these objects the most desired object category is "home & kitchen" and the most common use for 3D printing at home is to "replicate existing objects" [13]. Examples for objects in these categories are fruit bowl, small dishes, teapot or coffee mug [13]. In this study the focus is on creating a (fruit) bowl. This object seems to be designable for non-experts but at the same time designing this object in the respective toolkit might be a challenge for non-experts. There is no specification on how to create this bowl. Participants are just supposed to give their best on creating this item. Before the study starts an example of such a created bowl is shown to every participant (Figure 3).

Figure 3: Example of a created bowl in the toolkit

With reference to the participants, the target group are students who do not have any designing experience. However, these students are supposed to have a bit of technical affinity in order to be able to create an object. The majority of the students are students at the technical faculty of the University Erlangen-Nuremberg but students who haven't attended any computer-aided design course yet (non-experts with technical affinity).

As far as the toolkit is concerned, Autodesk 123D is chosen as the toolkit for the study. There is a huge variety of tools. As the study targets people without designing skills tools for beginners are focused. Autodesk 123D is recommended for beginners [shapeways] but it has many functionalities in order to create simple objects. Furthermore this tool is for free (freemium).

After the study, participants are asked to report their opinion about the toolkit and about 3D printing in general. Seven-position categorical scales with boxes labeled from likely (extremely, quite, slightly) to unlikely (extremely, quite, slightly) are provided in a questionnaire (cf. appendix: Figure 20, Figure 21, Figure 22).

To every participant this study is introduced as follows: "Hi. Are you interested in a quick survey/study?" "Yes." "Ok, I want to find out what you think about a specific 3D printing toolkit and what you think about 3D printing in general. Basically what you need to do is to create a simple object in a toolkit, called Autosdesk 123D, and after you have to answer a one page questionnaire. Certainly I am going to support you with the basic functionalities (tools) of that toolkit. You cannot make anything wrong. Every result is valuable for my research. The object I am talking about is a bowl. An example of a bowl could look like that. So now I just hand the mouse over to you and please just try to get started with the toolkit."

4   Data analysis

The total number of participants is 30. The age of these participants is shown in Figure 4. Thereby the average age is 24 years with the majority of students being between 24 and 25 years old.

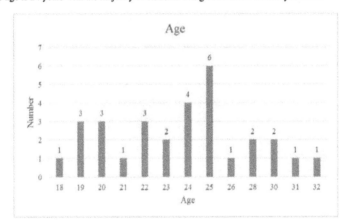

Figure 4: Age of participants

Figure 20 shows the distribution of the age. The proportion of female participants is 10 %.

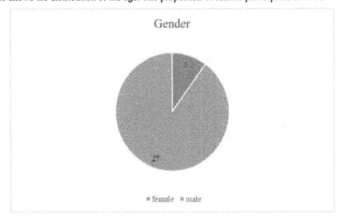

Figure 5: Gender of participants

The participating students are part of the following studies: Communication and Multimedia Engineering, Energy Management, Economics, Trade Management, Computational Engineering, Industrial Engineering, Biology, Electrical Engineering, Medical Engineering, Chemistry, Nanotechnology and Physics.

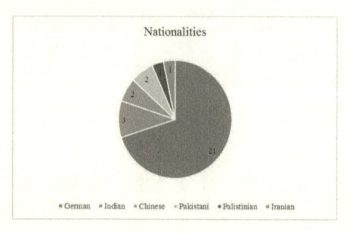

Figure 6: Nationalities of participants

Figure 6 shows the distribution of the nationalities. The majority of participants are Germans followed by Indians.

	Mean	Standard Deviation
It took me X minutes to create the object: [minutes]	11,33	4,34
Self-efficacy		
I could complete a task using the toolkit…		
a) if there was no one around to tell me what to do as I go.	3,93	1,80
b) if I had a lot of time to complete the job for which the toolkit was provided.	2,73	1,57
Anxiety		
I feel apprehensive about using the toolkit.	5,60	1,38
It scares me to think that I could lose a lot of information using the toolkit by hitting the wrong key.	5,93	1,31
I hesitate to use the toolkit for fear of making mistakes I cannot correct.	5,67	1,18
Attitude towards this technology (Usefulness)		
I find the toolkit useful.	2,73	1,34
Using the toolkit enhances my effectiveness.	3,87	0,68
The toolkit supports to get my ideas into physical objects.	3,93	1,55
Quality Output		
The quality output I get from the toolkit is high.	4,03	1,38
I have no problem with the quality of the toolkit's output.	4,27	1,48
Image		
Having the toolkit or a 3D printer is a status symbol.	4,73	1,23
People who use the toolkit or a 3D printer have more prestige than those who do not.	4,87	1,25
Social influence		
People who influence my behavior think that I should use the toolkit or 3D printing.	4,73	1,23
People who are important to me think that I should use the toolkit or 3D printing.	4,87	1,25
Usability (Ease of Use)		
Learning to deal with the toolkit was easy.	3,73	1,64
I find it easy to achieve with the system exactly what I want.	4,63	1,27
My interaction with the toolkit was clear and understandable.	4,13	1,53
It would be easy for me to become skillful at using the toolkit.	2,83	1,09
Behavioral Intention to Use (Usage)		
I intend to use such a toolkit in the next 12 months.	5,70	1,70
I am sure I am going to use a 3D printing toolkit in the upcoming years.	3,57	1,68

Table 1: Measurement scales and scale items with mean value and standard deviation

Table 1 displays the compressed results of the study. The average time to create the fruit bowl is 11,3 minutes. In this time people tried to create a bowl already having an example of a bowl in mind as an example was presented to every participant while introducing this study (Figure 3). It can be assumed that the result of the designing would not have been better even if the participants would have taken more time to create a bowl. For further designing improvements tutorials and other resources might have been taken into account. The highest mean value of all measurement scales with 5,7 is the intention to use such a toolkit in the next 12 months. So, people are most likely not to use such a tool in the next months. However, the majority of students was not able to create something similar to a bowl. Designing examples are displayed in Figure 7.

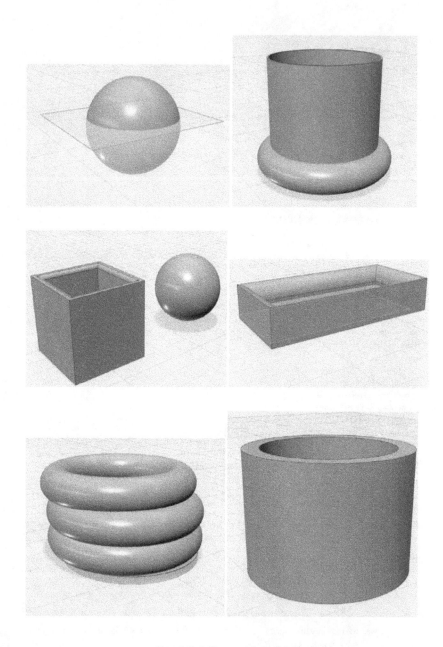

Figure 7: Designing examples of participants

Self-efficacy

Having a closer look at scale items many students were optimistic, even though they were not able to create a bowl in the set time, to complete the task if there was no one around them. In everyday life people would use tutorials and other sources in order to learn about the tool. With these sources the success rate might have been higher. On the other hand 13 students have just little confidence to complete this task even with tutorials etcetera. The mean value for this measurement scale is 3,93 with categorical scales from one (extremely likely) to seven (extremely unlikely).

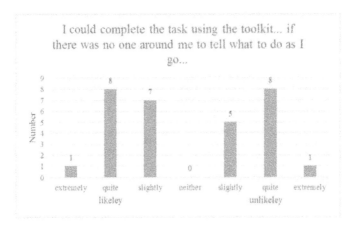

Figure 8: Scale item: I could complete a task using the toolkit if there was no one around to tell me what to do as I go.

Asking the participants whether they are able to finish a bowl with unlimited time 8 of 30 predict to be successful to create this object without time limit. The mean value is 2,73. Consequently, many students are confident about finishing this task when they do not have any time pressure and if they have freedom for working into this new task (designing) which most of the students have never done before.

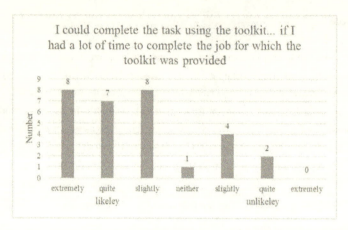

Figure 9: Scale item: I could complete a task using the toolkit if I had a lot of time to complete the job for which the toolkit was provided.

Anxiety

The following was explained to every single interviewee in order to explain the goal of the questions of this category: "Some people feel apprehensive etcetera when they see a lot of buttons in a tool or software which they might have never seen before. So how did you feel by using this tool?"

Out of the three scale items of the measurement scale anxiety one scale item is presented in Figure 10. The result of all three scale items are similar. Nevertheless, even though most of the interviewed students have never worked with any Computer Aided Design software before, they are neither scared, apprehensive nor hesitant. 20 out of the 30 respondents are stating that they are quite/extremely unlikely hesitating to use this toolkit. The mean value is 5,67.

Figure 10: Scale item: I hesitate to use the toolkit for fear of making mistakes I cannot correct.

Attitude towards this technology (Usefulness)

This measurement scale is supposed to display whether participants think that Autodesk 123D for creating simple 3D objects is useful and whether this toolkits is able to transfer individual's thoughts into digital objects. While the majority of interviewees think that this toolkit is useful (mean value of 2,73) there are some students who do not think so. One reason is that people without designing skills do not really have a chance to create a simple object such as a bowl. This tool is nice to create something but the more specific the idea becomes the less non-experts might be in a position to get these ideas into the tool.

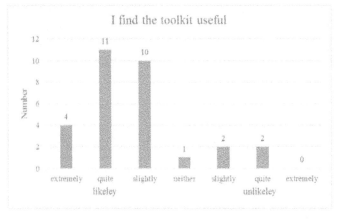

Figure 11: Scale item: I find the toolkit useful.

Consequently, asking more concrete, whether the toolkit supports to get ideas into physical objects, the result is different (Figure 12). The mean value is 3,93, so much higher than the preceding mean value. As the mean value is >3,5 it can be followed that it is more unlikely than likely that the toolkit supports to get ideas into physical objects.

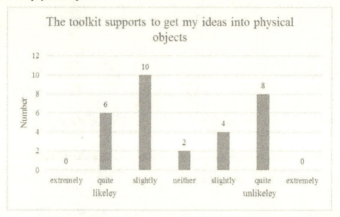

Figure 12: Scale item: The toolkit supports to get my ideas into physical objects.

Quality Output

This measurement scale consists of two scale items. One scale item is presented in Figure 13. The mean value is 4,03. Even though the participants did not really get the output they wanted (bowl) the rating is not really on the "unlikely" side. One reason can be seen in the optimistic rating behavior of the students. Many of these students have never worked with any designing tool before. Hence, they were surprised with the result they achieved in these ten minutes although they did not design the required object. However, a difference between nationalities could be observed as well. While especially Germans are pretty conservative in their rating (rating rather unlikely than likely) Asians are much more optimistic (rating rather likely than unlikely). However, one result of this data is that the used toolkit is not easy enough in order to enable people to design easy household objects.

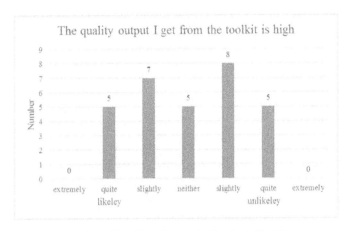

Figure 13: Scale item: The quality output I get from the toolkit is high.

Image and Social influence

The measurement scales image and social influence and their respective scale items are pointing in a similar direction. The data results are similar as well. Therefore, one of these four scale items is shown in Figure 14. The mean value is 4,73. Most of the participants think that a 3D printer (with a toolkit) is rather not a status symbol and that people with such technology do not necessarily have more prestige than those who do not own such technology. Moreover, the social influence from people who are important to the interviewees to use this technology is rather small. However, there are some people who think that this technology is prestigious and that in a couple of years the image of this technology (status symbol) might play a more crucial role.

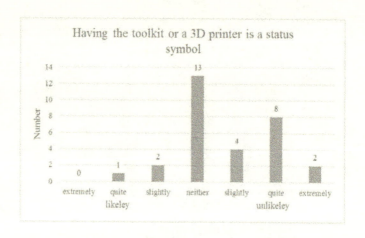

Figure 14: Scale item: Having the toolkit or a 3D printer is a status symbol.

Usability (Ease of Use)

Ease of Use refers to "the degree to which a person believes that using a particular system would be free of effort" [12]. This measurement scale consists of four scale items. Two scale items are selected and displayed in the following.

The mean value in Figure 15 is 4,63. Hence, the majority stated that the toolkit did not really transfer the idea into a digital design.

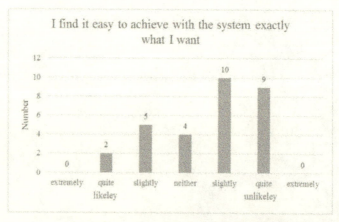

Figure 15: Scale item: I find it easy to achieve with the system exactly what I want.

However, most of the interviewees think that they would be able to become skillful with this toolkit just by trial and error. The mean value is 2,83.

Figure 16: Scale item: It would be easy for me to become skillful at using the toolkit.

Behavioral Intention to Use (Usage)

Participants were asked to report their intention to use a toolkit or 3D printing in the next 12 months and in the upcoming years. With a mean value of 5,7 many students predict not to use such a toolkit in the next 12 months. Many of them explained that this is simply because of they would not need it in their studies. Nonetheless, when 3D printing spreads further and these tools might become easier some of the students can imagine to use a toolkit in the near future.

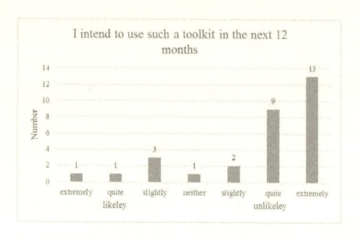

Figure 17: Scale item: I intend to use such a toolkit in the next 12 months.

By asking for a prediction whether participants might use 3D printing and a toolkit in the upcoming years, the mean value is 3,57 (Figure 18). So many students could imagine themselves to get in touch with 3D printing within the next years. One lady stated: "It would be great just to print out my own customized fork and knife and when I have kids I just could replace easily every cutlery my kids might destroy."

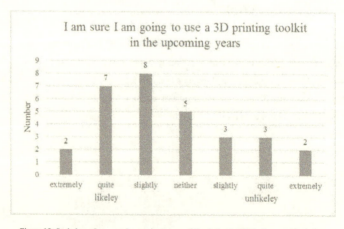

Figure 18: Scale item: I am sure I am going to use a 3D printing toolkit in the upcoming years.

5   Correlation statistic

In this section a reliability analysis will be executed by calculating Cronbach alpha. Calculating Cronbach alpha is usually done when TAM is applied. Cronbach alpha indicates the internal consistency between the different scale items of one measurement scale [11]. High internal consistency is given when the respective scale items correlate high, so they are very similar to each other [11]. Alpha can reach a maximum level of +1, the smaller the value the lower the consistency [11].

Cronbach alpha for the measurement scales of this study is calculated as follows:

$$\alpha = \frac{N*\bar{r}}{1+(N-1)*\bar{r}}$$

Results:

Anxiety: 0,79

Self-efficacy: 0,81

Attitude toward this technology (Usefulness):   0,49

Quality Output: 0,74

Image: 0,89

Social influence: 0,89

Usability (Ease of Use):   0,83

Behavioral Intention to Use (Usage):   0,58

The process of calculation is part of the appendix (Table 3). Self-efficacy, Image, Social influence and usability do have an excellent internal consistency. The lowest value has Usefulness: One reason is that the scale item "using the toolkit enhances my effectiveness" has been answered in most of the interviews with "neither". This question should have been eliminated prior to the survey due to the fact that people are not able to answer this question when they are not in a position to finish the target object. Consequently, internal consistency of this measurement scale drops.

The proposed research model for the usage of 3D printing toolkits was introduced in Figure 2. This model is supposed to show the relations between determinants, such as self-efficacy and social influence, and the Behavioral Intention to Use a 3D printing toolkit. For each determinant (measurement scale) at least two scale items have been developed in order to calculate and evaluate the influence and relationship of determinants on the Intention to Use. For the calculation the Pearson correlation coefficients are calculated. Table 2 shows the results. The correlation coefficients can always reach a value from -1 to +1. The higher the absolute value the higher the relationship. Effect sizes are defined as follows [11]: Low effect sizes for an absolute value <0,15, middle effect sizes for an absolute value from 0,15-0,35, high effect sizes for absolute values >0,35.

	Behavioral Intention to Use (Usage)
Self-efficacy	
I could complete a task using the toolkit…	
a) if there was no one around to tell me what to do as I go.	0,63
b) if I had a lot of time to complete the job for which the toolkit was provided.	0,66
Anxiety	
I feel apprehensive about using the toolkit.	-0,54
It scares me to think that I could lose a lot of information using the toolkit by hitting the wrong key.	-0,55
I hesitate to use the toolkit for fear of making mistakes I cannot correct.	-0,27
Attitude towards this technology (Usefulness)	
I find the toolkit useful.	0,17
Using the toolkit enhances my effectiveness.	0,27
The toolkit supports to get my ideas into physical objects.	0,55
Quality Output	
The quality output I get from the toolkit is high.	0,64
I have no problem with the quality of the toolkit's output.	0,40
Image	
Having the toolkit or a 3D printer is a status symbol.	0,09
People who use the toolkit or a 3D printer have more prestige than those who do not.	0,20
Social influence	
People who influence my behavior think that I should use the toolkit or 3D printing.	0,09
People who are important to me think that I should use the toolkit or 3D printing.	0,20
Usability (Ease of Use)	
Learning to deal with the toolkit was easy.	0,81
I find it easy to achieve with the system exactly what I want.	0,83
My interaction with the toolkit was clear and understandable.	0,90
It would be easy for me to become skillful at using the toolkit.	0,64
Behavioral Intention to Use (Usage)	
I intend to use such a toolkit in the next 12 months.	-
I am sure I am going to use a 3D printing toolkit in the upcoming years.	-

Table 2: Correlation matrix

6 Evaluation

In this section the results of the data analysis are discussed and evaluated. This study examines the use of 3D printing toolkits. Interpretation and findings are detailed below.

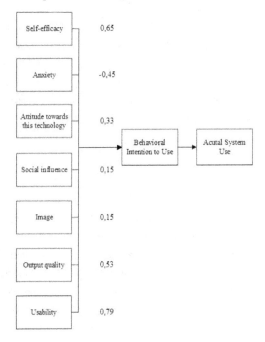

Figure 19: Results of data analysis: Measurement scale correlation with Behavioral Intention to Use

Figure 19 shows the correlations of determinants and the Behavioral Intention to Use a 3D printing toolkit. For each determinant (measurement scale) the average of the respective scale items has been calculated.

Between *self-efficacy* and the Behavioral Intention to Use a high positive effect size can be stated (0,65). It is quite obvious that the higher the self-efficacy of a student the more likely she/he is going to use such a toolkit. When someone does not believe she/he would be able to complete a job she/he in consequence does not have fun in this task and she/he might avoid this technology in future. However, self-efficacy does strongly affect the Behavioral Intention to Use of non-experts.

Anxiety and Behavioral Intention to Use correlate negative (-0,45). This is a quite strong negative correlation. How can this be interpreted? This means that the higher the anxiety the lower the Behavioral intention to Use. Hence, anxiety does strongly affect the Behavioral Intention and if someone is apprehensive about this toolkit he probably won't use this toolkit and 3D printing technology in future.

The *attitude towards this technology* has an effect size of 0,33 which is a middle effect size. The more useful the toolkit is and the better it transfers ideas into physical objects the higher the Behavioral Intention to Use the toolkit might be. Hence, making toolkits more intuitive and easier will directly influence the Usage of these toolkits.

Image and Social influence have a low effect size on the Behavioral Intention. Following, the interviewees do not think that image and social influence do have strong influence on their usage. However, there is little relationship between these determinants and the Behavioral Intention.

Not least *Quality Output* and *Usability (Ease of Use)* both have strong influence on the Behavioral Intention. The higher the quality of the toolkit's output and the easier the toolkit for people without designing skills the more likely people are going to use these.

7 Conclusion

The purpose of this study was to develop and validate a research model which predicts the Actual System Use of 3D printing toolkits. Seven measurement scales have been hypothesized which are supposed to be determinants of 3D printing toolkits usage. Cronbach alpha confirmed high internal consistencies for each measurement scale, except of *Attitude towards this technology (Usage)*, as discussed. A correlation matrix stated that five out of these seven determinants do strongly affect the Behavioral Intention to use a 3D printing toolkit. However, *Image and Social Influence* do have weak a relation to the Behavioral Intention to Use. Therefore, these five measurement scales, which do have strong influence should be focused on in order to make 3D printing in general more attractive to non-experts. The author thinks that three measurement scales (*Attitude towards this technology, Quality Output, Usability*) do directly address the toolkits and the other remaining two determinants (*self-efficacy, anxiety*) are determinants which differ from person to person. Software corporations such as Autodesk need to focus on these three measurement scales which belong to the toolkits itself and the other two determinants which are individual related might be influenced by supporting people. One example of support could be more online-tutorials or the opportunity of live-chats which could be used by hobby designers when designing

References

[1] B. McKenna, *"Wohlers Report 2014" Is Out! The Top 3 Takeaways From the Gold Standard on the 3-D Printing Industry"*. Available: http://www.fool.com/investing/general/2014/05/05/wohlers-report-2014-is-out-the-top-3-takeaways-fro.aspx.

[2] J. Becker and D. Pfeiffer, "Beziehungen zwischen behavioristischer und konstruktionsorientierter Forschung in der Wirtschaftsinformatik," in *Fortschritt in den Wirtschaftswissenschaften*, S. Zelewski and N. Akca, Eds, Wiesbaden: DUV, 2006, pp. 1–17.

[3] A. R. Hevner, S. T. March, J. Park, and S. Ram, "Design Science in Information Systems Research," *MIS Quarterly*, vol. 28, no. 1, pp. 75–105, 2004.

[4] F. D. Davis, "A technology acceptance model for empirically testing new end-user information systems : theory and results," Dissertation, Sloan School of Management, Massachusetts Institute of Technology, Massachusetts, 1986.

[5] J. Kotrík, "UTAUT - die gegenwärtige Weiterentwicklung von TAM," term paper, Institut für Informationswirtschaft, University of Economics and Business Vienna, Vienna, 2009.

[6] M. Chuttur, "Overview of the Technology Acceptance Model: Origins, Developments and Future Directions," *Sprouts: Working Papers on Information Systems*, vol. 9, no. 37, pp. 1–21, http://sprouts.aisnet.org/785/1/TAMReview.pdf, 2009.

[7] F. D. Davis, R. P. Bagozzi, and P. R. Warshaw, "User Acceptance of Computer Technology: A Comparison of Two Theoretical Models," *Management Science*, vol. 35, no. 8, pp. 982–1003, 1989.

[8] V. Venkatesh and F. D. Davis, "A Theoretical Extension of the Technology Acceptance Model: Four Longitudinal Field Studies," *Management Science*, vol. 46, no. 2, pp. 186–204, 2000.

[9] V. Venkatesh, F. D. Davis, G. B. Davis, and M. G. Morris, "User Acceptance of Information Technology: Toward a Unified View," *MIS Quarterly*, vol. 27, no. 3, pp. 425–478, 2003.

[10] V. Venkatesh and H. Bala, "Technology Acceptance Model 3 and a Research Agenda on Interventions," *Decision Sciences*, vol. 39, no. 2, pp. 273–315, 2008.

[11] P. Traxler, *Die Bedeutung von Einstellung und Motivation für den Einsatz von E-Learning durch Lehrende an Pädagogischen Hochschulen*. Glückstadt: Hülsbusch, 2013.

[12] F. D. Davis, "Perceived Usefulness, Perceived Ease of Use, and User Acceptance of Information Technology," *MIS Quarterly*, vol. 13, no. 3, pp. 319–340, http://www.jstor.org/stable/249008, 1989.

[13] R. Shewbridge, A. Hurst, and S.K. Kane, Eds, *Everyday making: identifying future uses for 3D printing in the home*: ACM, 2014.

8 Appendix

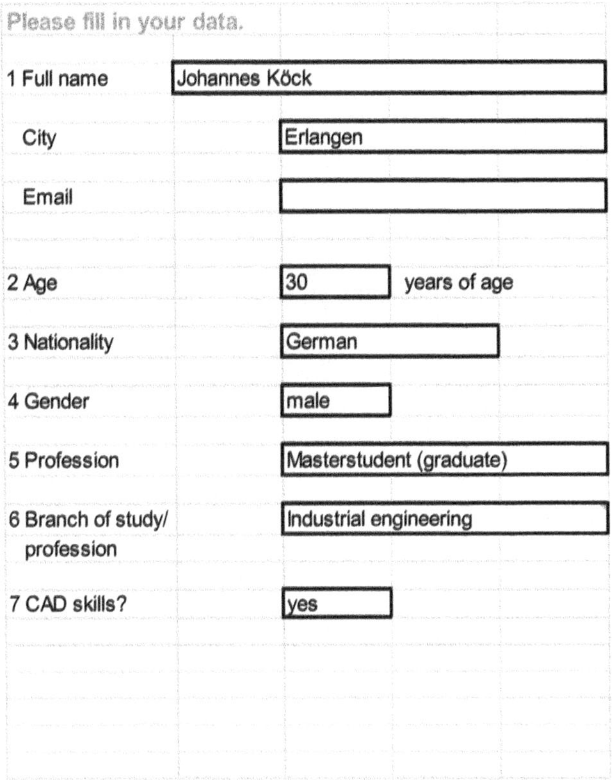

Please fill in your data.

1 Full name	Johannes Köck
City	Erlangen
Email	
2 Age	30 years of age
3 Nationality	German
4 Gender	male
5 Profession	Masterstudent (graduate)
6 Branch of study/ profession	Industrial engineering
7 CAD skills?	yes

Figure 20: Page 1 Questionnaire: General data of participants

Background information: "With this study I want to find out whether non-experts are able to design easy object in a 3D printing toolkit."

Instructions for study

1. Please download and install the free software to create 3D models: http://www.123dapp.com/design

2. Please create a (fruit) BOWL according to your fantasy. In the table below there is an example of mine. There are many ways to create such a bowl in Autodesk 123D.

3. Please note: Please be patient, this task is not easy at all even though this is one of the easiest objects to create. It might take some time.

4. If you cannot finish the model, that is fine too. This would be a result for the study as well.

5. You are also allowed to use youtube tutorials.

6. If you have any questions let me know asap.

7. After, please fill out "questions". This is the most important part for the study.

8. In the end pls. send this filled sheet and your * .123dx file to me (johannes@vanyon.de).

Figure 21: Page 2 Questionnaire: Instruction for study

29

It took me X minutes to create the object:

Self-efficacy

I could complete a task using the toolkit…

a) if there was no one around to tell me what to do as I go.

likely								unlikely
	extremely	quite	slightly	neither	slightly	quite	extremely	

b) if I had a lot of time to complete the job for which the toolkit was provided.

likely								unlikely
	extremely	quite	slightly	neither	slightly	quite	extremely	

Anxiety

I feel apprehensive about using the toolkit.

likely								unlikely
	extremely	quite	slightly	neither	slightly	quite	extremely	

It scares me to think that I could lose a lot of information using the toolkit by hitting the wrong key.

likely								unlikely
	extremely	quite	slightly	neither	slightly	quite	extremely	

I hesitate to use the toolkit for fear of making mistakes I cannot correct.

likely								unlikely
	extremely	quite	slightly	neither	slightly	quite	extremely	

Attitude towards this technology (Usefulness)

I find the toolkit useful.

likely								unlikely
	extremely	quite	slightly	neither	slightly	quite	extremely	

Using the toolkit enhances my effectiveness.

likely								unlikely
	extremely	quite	slightly	neither	slightly	quite	extremely	

The toolkit supports to get my ideas into physical objects.

likely								unlikely
	extremely	quite	slightly	neither	slightly	quite	extremely	

Quality Output

The quality output I get from the toolkit is high.

likely								unlikely
	extremely	quite	slightly	neither	slightly	quite	extremely

I have no problem with the quality of the toolkit's output.

likely								unlikely
	extremely	quite	slightly	neither	slightly	quite	extremely

Image

Having the toolkit or a 3D printer is a status symbol.

likely								unlikely
	extremely	quite	slightly	neither	slightly	quite	extremely

People who use the toolkit or a 3D printer have more prestige than those who do not.

likely								unlikely
	extremely	quite	slightly	neither	slightly	quite	extremely

Social influence

People who influence my behavior think that I should use the toolkit or 3D printing.

likely								unlikely
	extremely	quite	slightly	neither	slightly	quite	extremely

People who are important to me think that I should use the toolkit or 3D printing.

likely								unlikely
	extremely	quite	slightly	neither	slightly	quite	extremely

Usability (Ease of Use)

Learning to deal with the toolkit was easy.

likely								unlikely
	extremely	quite	slightly	neither	slightly	quite	extremely

I find it easy to achieve with the system exactly what I want.

likely								unlikely
	extremely	quite	slightly	neither	slightly	quite	extremely

My interaction with the toolkit was clear and understandable.

likely								unlikely
	extremely	quite	slightly	neither	slightly	quite	extremely

It would be easy for me to become skilful at using the toolkit.

likely								unlikely
	extremely	quite	slightly	neither	slightly	quite	extremely

Behavioral Intention to Use (Usage)

I intend to use such a toolkit in the next 12 months.

likely								unlikely
	extremely	quite	slightly	neither	slightly	quite	extremely

I am sure I am going to use a 3D printing toolkit in the upcoming years.

likely								unlikely
	extremely	quite	slightly	neither	slightly	quite	extremely

Figure 22: Page 3 Questionnaire: Measurement scales and items

Anxiety				r^2	
	1	2	3		
1	1	0,6327955	0,464337602	0,560632701	**0,7928749**
2		1	0,584765011		
3			1		

Self-efficacy			r^2	
	1	2		
1	1	0,6753247	0,675324675	**0,8062016**
2		1		

Attitude towards this technology (Usefulness)				r^2	
	1	2	3		
1	1	0,3380176	0,140647909	0,24359408	**0,4913852**
2		1	0,252116729		
3			1		

Quality Output			r^2	
	1	2		
1	1	0,5862187	0,58621866	**0,7391398**
2		1		

Image			r^2	
	1	2		
1	1	0,8046067	0,804606675	**0,8917253**
2		1		

Social influence			r^2	
	1	2		
1	1	0,8046067	0,804606675	**0,8917253**
2		1		

Usability (Ease of Use)				r^2	
	1	2	3		
1	1	0,5137171	0,594183846	0,626692626	**0,8343349**
2		1	0,772176963		
3			1		

Behavioral Intention to Use (Usage)			r^2	
	1	2		
1	1	0,4117346	0,411734567	**0,5833031**
2		1		

Table 3: Process of calculation: Cronbach alpha